D0200000

101 Questions
About the Seashore

Sy Barlowe

DOVER PUBLICATIONS, INC.
Mineola, New York

Bibliographical Note

101 Questions About the Seashore is a new work, first published by Dover Publications, Inc., in 1997.

Library of Congress Cataloging-in-Publication Data

Barlowe, Sy.
   101 questions about the seashore / Sy Barlowe.
      p.   cm.
  ISBN 0-486-29914-7 (pbk.)
  1. Seashore—Miscellanea.   I. Title.
GB451.2.B35   1997
577.5'1—dc21   97-20552
                                         CIP

Manufactured in the United States of America
Dover Publications, Inc., 31 East 2nd Street, Mineola, N. Y. 11501

Table of Contents

I. Introduction

1. Why do we study the life of the shore and shallow sea?
Almost three fourths of the earth is covered by water, in
which life had its origin. We all carry that heritage in the salt
content of our blood. Along the edges of the oceans and seas
lie narrow bands of land, seashores, which are home to an
astounding variety of life forms. Entire species of animals
and plants are largely governed by the ebb and flow of the
sea: barnacles, starfish, sea anemones, and seaweed, whose
unique life cycles depend on their exposure to the air for at
least an hour or two every day. Others thrive beyond the
immediate influence of the tides, amidst the eelgrass and seaside plants. The survival and well-being of the inhabitants of
the seashores depend on the vagaries of daily and seasonal
changes in temperature, light, and humidity.

The kinds of animals that are found at the shore vary with
the kinds of coastline and climate. The life of the rocky coast
of Maine differs from that of the sandy shores of Long
Island, and that of tropical regions differs from both.

The shoreline is a dynamic entity, constantly changed by
the action of the wind and waves. Storm waves erode beach

1

cliffs, carrying away tons of sand to be deposited elsewhere and creating new land forms in the endless battle between sea and shore. An awareness of this interplay of forces is essential to preserving the environmental integrity of these vital areas.

2. What kinds of shore habitats are there? The most familiar area is the sandy beach, the least productive habitat for animals and plants because of the crashing surf and shifting sands. Here the only life is hidden beneath the surface in burrows. Rocky shores present a more interesting habitat, providing anchors for seaweed, barnacles, anemones, and snails and hiding places for crabs, starfish, and sea urchins. When the tide recedes, it leaves tide pools, variously sized puddles of seawater among the rocks. These pools are filled with a great variety of living creatures. Mud flats and estuaries are yet another environment teeming with life forms.

3. How does the sea affect our climate? As the primary source of the atmospheric moisture that becomes rainfall, the oceans provide the freshwater necessary for life on earth. The sun warms the sea, causing evaporation; as the vapor condenses, it creates clouds and, eventually, rain. The rain finds its way to rivers, where it completes the cycle by returning to the sea. Inland temperature variations are caused in part by winds that have been warmed or cooled by ocean currents.

2. General

4. Why is the sea salty? Billions of years ago the ocean waters were almost salt-free. As the seawater evaporated, formed clouds, turned into rain, and poured onto the land, creating lakes and rivers, the consequent runoff contained dissolved salts, the accumulation of which gave us our saltwater seas and oceans. Today, a cubic mile of sea water is estimated to be 3.5 percent salt and contains some 166 million tons of salt.

5. What are the most important minerals in seawater? Since supplies of minerals on land are declining, it is inevitable that we will have to turn to the sea for these valuable resources. In order of the quantities that can be extracted from 1 cubic mile of seawater, they are sodium chloride, magnesium chloride, magnesium sulfate, calcium sulfate, potassium sulfate, and calcium carbonate.

6. What do marine animals eat? They feed on plankton, seaweed, and one another, using a variety of methods: extending sticky tentacles, ingesting microscopic organisms, browsing on algae, and capturing prey by stealth.

7. What are mollusks? Mollusks are a group of about 100,000 species of mostly marine animals, including clams, snails, squid, and octopuses. Marine mollusks are to be found at all depths of the ocean and even floating on the surface. They live on and in the sand, among and under rocks.

The shells of the various kinds of mollusks are so distinctively beautiful that they appear in many museums and private collections. They are also used as jewelry, ornaments and buttons.

8. What is a radula? Many mollusks are equipped with a rasping organ called a radula, a ribbonlike structure covered on one side with many rows of very small teeth that allow the

Radula

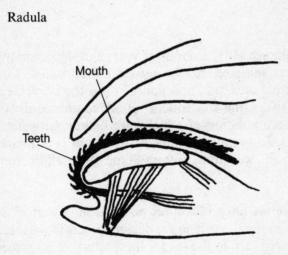

animal to scrape algae or drill through the shell of some unfortunate oyster or clam. The teeth of the radula point backward to draw the food back into the mollusk's throat.

9. What are bivalves? These are a group of mollusks also named pelecypods, Greek for "hatchet-footed." Bivalves are

Bivalves

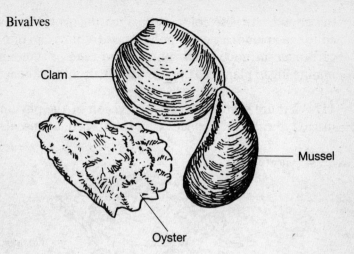

Clam

Mussel

Oyster

the only mollusks with two hinged shells, or valves, that can be opened or closed by powerful sets of muscles. Examples of bivalves are clams, oysters, and mussels.

10. Are corals confined to tropical seas? No. Although they are found mainly in the warm waters of the tropics, a few occur in temperate regions, and some even thrive in the Arctic. Thousands of coral islands called atolls are the result of

Atoll

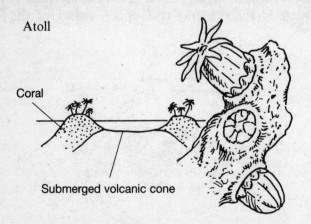

Coral

Submerged volcanic cone

the growth of these colonies through the ages. Closely related to sea-anemones, coral polyps build a hard cup of calcium carbonate in uncounted billions and take on the colors of minute algae plant cells present for their mutual benefit.

11. What are gastropods? A subdivision of the phylum Mollusca, typically they consist of a spiral shell, a muscular foot,

Gastropod

Operculum

a head with eyes, tentacles, and an operculum (see 12). Gastropods make up the largest group of mollusks and include land forms as well. Some typical examples are whelks, abalones, cone shells, and periwinkles.

12. What is an operculum? Found in most snail species, it is a horny plate attached to the rear upper surface of the foot.

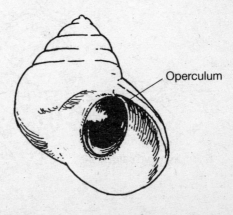

Operculum

When the foot is withdrawn into the shell, this "door" snugly fits the opening and seals shut, protecting the snail from its enemies. The operculum also prevents some snails from drying out when the tide recedes. One of the tropical snails produces an operculum that is known as the cat's eye, so beautiful that it is used in jewelry.

13. Why is the sea blue? The blue color of the sea results from the absorption and reflection of light by water molecules, debris and dust, and the amount of plankton present. Light rays passing through the water are subject to scattering by these elements, with red, yellow, and green rays being diffused more than the blue. The fewer particles there are in the sea to reflect these rays, the more intense the blue will be. Moreover, the ocean reflects the sky. Coastal waters differ because there are more particles and plankton to absorb the light, and they reflect more colors. The concentration of phytoplankton in these waters produces chlorophyll, which imparts a greenish hue to the water.

14. What are arthropods? The phylum Arthropoda, which means "jointed-footed" in Greek, includes crustaceans, horseshoe crabs, sea spiders, and insects. This book focuses on the marine species, such as crabs, lobsters, shrimps, and sand fleas. As arthropods develop, they cast off their rigid outer covering, revealing a soft new one that soon hardens into a larger replica of the old one. They are bilaterally symmetrical: each side of the body mirrors the other.

15. What is the exoskeleton? It is the rigid outside covering (external skeleton) that protects and supports muscles, allowing movement of arthropods. This armorlike material periodically splits, allowing the growing animal to molt by shedding the older exoskeleton. During the molting, the new covering is soft, and the animal is vulnerable to attack by predators. The most familiar examples are the lobster and the blue crab.

16. What are some examples of crustaceans? The phylum Crustacea, named for the animals' "crust," or outer covering, refers to some 30,000 joint-legged species living in the sea, in freshwater, and even on land. These include crabs, lobsters, shrimps, beach fleas, wood lice, barnacles, and water fleas. They can be sedentary, parasitic, or very active and have in common a segmented exoskeleton jointed at various places to allow freedom of movement. Crustaceans range in size from microscopic to very large lobsters to the giant king crab of Japan.

17. What are cilia? Cilia are minute internal or external hairs in many sea creatures. They vibrate, providing locomotion, and sweep currents containing food into the mouth.

Trochophore (larva of mollusks, annelids, etc.)

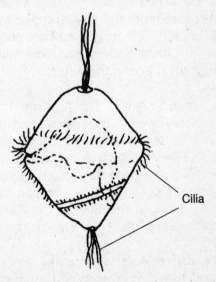

Cilia

18. What are echinoderms? Echinoderms, from the Greek echinos (hedgehog) and derma (skin), are animals whose external skeleton is made up of calcareous plates covered with projecting spines. They usually have a five-part, starlike pattern. There are five classes in this group: starfishes, sea urchins, brittle stars, sea cucumbers, and crinoids, or sea lilies.

19. What is symbiosis? There are relationships between animals of different species living together that scientists call symbiosis ("living together" in Greek). If the relationship is so vital that one cannot survive without the other, it is known as mutualism. When the intimate association benefits both but is not necessary for either, it is called commensalism. The hermit crab, using an empty shell, is commensal with the sea anemone found on the shell. The crab provides food for the anemone by stirring up the bottom in its search for food, while the anemone affords some protection to the crab by virtue of its stinging tentacles. If one animal causes harm to the other, the relationship is called parasitism.

Symbiosis

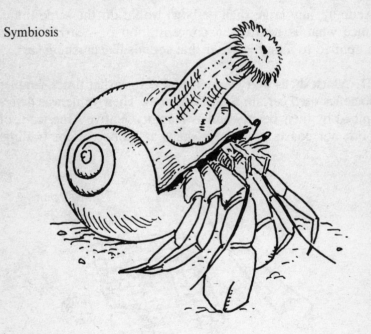

20. Which shell is commonly believed to reproduce the "roar of the ocean" when held to the ear? The queen conch (Strombus gigas) has fascinated children for generations with its supposed ability to bring the sound of the ocean to their ears.

Queen conch

Actually, any large shell or bowl would do the same thing, since what is heard is a concentration of nearby sounds amplified to a muffled roar that sounds like crashing surf.

21. Which shells were used as money? In earlier times, simpler societies used certain shells as money. Their value was determined by their beauty and availability. Native Americans of both our coasts used these shells as money or for trading.

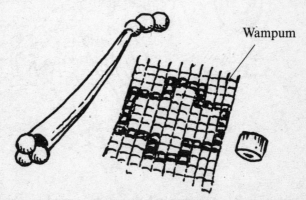

Wampum

The East Coast provided the quahog, whose shell was fashioned into beads and strung into strands or belts called wampum. The West Coast yielded the tooth shell (Dentalium), the shell of commerce. Traders collected cowrie shells in the Pacific and Indian oceans and traded them for valuable wares in Africa.

22. What are spring tides? Tides are governed by the rotation of the earth and the gravitational pull of the sun and moon. As the moon orbits the earth, twice a month its position is such that it is lined up with the earth and sun. These times are known as the full moon and new moon. The combined pull of the sun and moon create the highest and lowest tides, the spring tides.

23. What are neap tides? Twice a month the moon is at a right angle to the sun, somewhat neutralizing the latter's gravitational pull. At these times, known as the first and third quarter, the maximum rise and fall of the resulting tides, neap tides, are less than the high and low tides of the spring tides.

24. What is a tide pool? As the tide recedes, it exposes the green algae-covered rock formations hidden beneath the sea. Among the cracks and crevices revealed by the drop in water level are pools of water of varying dimensions. These are known as tide pools, home to a treasure trove of marine creatures that includes starfish, sea anemones, snails, sea urchins, and countless other inhabitants.

The occupants of these pools are a hardy lot since they are subject to variations in temperature, salinity, and tidal patterns.

Depressions on sandy shores, variations of the tide pool, are home to shrimps, crabs, and some small fish that have been stranded by the outgoing tide. Some of these depressions become permanent and contain a continuing living community.

25. What is sand made of? Sand arises from weathering and erosion. Each grain began as a rock or boulder. Because of fracturing by frost or tree roots and the incessant pounding of the waves, these rocks crumbled into smaller pieces. The constant grinding and buffeting by the sea reduced these pieces into tiny grains composed of quartz, feldspar, and other minerals found in the rocks. The whitish grains are the remains of ground-up clam and other shells; the smoother the grains, the older the sand.

26. What are the commercial uses of seaweed? Seaweed, or marine algae, is one of the most valuable products of our seacoasts. It is eaten in many countries, particularly in Japan, where cooks prepare many delicious dishes from it. Irish moss (Chondrus crispus) is eaten in Ireland, dulse (Rhodymenia palmata) in Scotland, and laver (Porphyra laciniata) in England and Wales. In the United States chemical and industrial products are extracted from seaweed. Mannitol, a derivative, is used in medicinal drugs and explosives. Other derivatives are found in food, candies, lotions, and manufacturing processes producing textiles, rubber, and other commodities.

27. What are Plankton? Plankton, from the Greek "drifters," consists of mostly microscopic single-celled plants (diatoms), and the larvae of most marine animals, along with tiny animals that spend their entire lives drifting with the currents of the open sea. The plants are called Phytoplankton, while the

Plankton

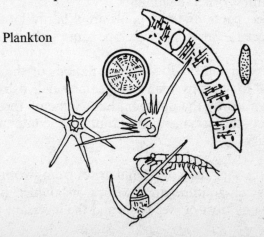

animal forms are known as Zooplankton. Though diatoms dominate in numbers, dinoflagellates, single-celled organisms possessing both plant and animal characteristics, are a very important element of the plankton. Some dinoflagellates are luminescent and emit a phosphorescent glow that is responsible for the sparkle of fiery light when oars are dipped into the water on a dark night. They are the cause of the deadly Red-Tide that on occasion kills thousands of fish. At other times, this drifting life occurs in such numbers that it colors the sea for miles around.

28. What are diatoms? They are microscopic plants that roam the seas in uncountable billions and provide the first link in the marine food chain. Essentially one-celled algae enclosed in intricate, glasslike containers of silica, they make up about 99 percent of the plant life of the sea. They come in a great variety of shapes: round, oval, square, and rodlike. They form an immense pasture, a grazing ground for swarms of copepods (see 76) and the larvae of many sea creatures. When diatoms die, they sink to the bottom and dominate the ocean floor, blanketing it for millions of square miles.

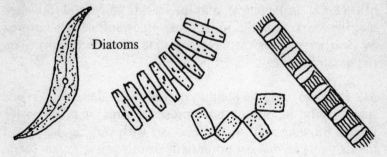

Diatoms

29. Is there a commercial use for diatoms? Yes. Vast areas of the sea bottom are covered by sediments known as oozes or red clay. Diatom ooze, consisting of silicified skeletons that have sunk to the ocean floor, has been mined for use by pharmaceutical companies as bacteria strainers. Other uses include strengthening concrete and producing a flat or semi-gloss finish in paint.

3. The Upper Seashore

30. What are some common seaside plants? Thriving in one of the most inhospitable environments in nature are sea rockets (*Cakile edentula*), poverty grass (*Hudsonia tomentosa*), beach peas (*Lathyrus japonica*), and the seaside goldenrod (*Solidago sempervirens*). In addition, stunted pines and cedars covered by bearberry vines display grotesque wind-razed forms on the harsh beaches.

31. How do seaside plants contend with damaging ocean spray? Wind at the seashore presents a major problem for plants. Blowing almost continuously with varying degrees of intensity, it can deposit potentially deadly salt spray on them. The successful survivors have adapted to this threat by means of several defensive measures: some have evolved leaves that present a high glossy surface that does not allow absorption; others have overlapping scales, like fish scales; still others have leaves that are covered with dense, tiny hairs that prevent the salt from coming into contact with the vulnerable surface.

32. How do trees adapt to wind and climatic conditions at the seashore? Unable to grow vertically because the windborne sand blasts and chisels branches unmercifully, trees and shrubs adapt by growing horizontally in pancake fashion. Twisted trunks and branches reach out 20 or 30 feet on a base that may only be 2 or 3 feet high.

Wind-shaped tree

33. What is poverty grass? As the wind blows sand, creating dunes, poverty grass (*Hudsonia tomentosa*), with its abundant, small, yellow flowers, is usually one of the first plants to take root. Over time, humus is created by the combined action of dead-leaf deposits and stems in combination with the deep roots. The resulting process provides an environment or soil that is capable of retaining water and allows larger and more demanding plants to flourish.

34. What shorebirds are common? Millions of birds rely on the sea for their survival. From the diving grebes of the Arctic to the penguins of the Antarctic, the birds that grace our seas and shores present a variety of beauty and form adapted

Tern

Heron

Curlew

Turnstone

Plover

to their unique needs. Some common inhabitants of the shoreline are gulls, herons, ducks, sandpipers, terns, sanderlings, plovers, oystercatchers, curlews, and avocets.

35. What bird is known as the fish hawk? The osprey (*Pandion haliaetus*) is called the fish hawk because its diet is exclusively fish. Hovering at heights up to 100 feet, it spots its prey, plunges with talons extended into the water, and grasps its slippery victim. The specially equipped feet of the osprey have spiny projections that prevent the scaly fish from escaping the deadly dive. It has been reported that some ospreys have drowned while attempting to lift too heavy a fish.

Osprey

36. What shorebirds are known as sea swallows? The swift, graceful, narrow-winged birds with deeply forked tails that flit about, diving or hovering along the shore, are terns, which are fondly referred to as sea swallows. The common tern (*Sterna hirundo*) breeds in large colonies. It defends its nest vigorously, attacking with its bill, sometimes with deadly results to small animals and birds.

37. Is the ghost crab real? Nocturnal and sand-colored, the swift-running ghost crab (*Ocypode quadrata*) is well named. At night this crab emerges from its 2- or 3-foot-deep burrow to feed on beach fleas and bits of refuse. It runs sideways on the tips of its toes, and when it stops, it seems to disappear by blending into the environment.

Beach flea

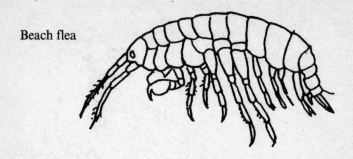

38. What are beach fleas? Feeding on dead plant and animal matter, these inch-long crustaceans are a source of food for shorebirds and occur in the water as well as on land.

39. Why are fiddler crabs called by that name? The male (*Uca pugilator*), not much larger than an inch wide, brandishes a fearsome, large claw that some have said resembles a violin; in motion, it suggests a violinist using his bow. This large pincer is, in fact, used in courtship and in battles with other males.

40. What do seagulls eat? Gulls, though found mainly along the seashore on all coasts in temperate regions, are also found

Seagull

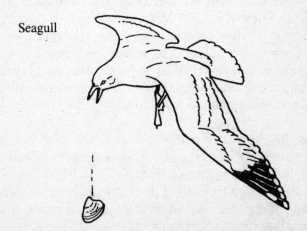

on marshes, lakes, and estuaries far inland. Although primarily scavengers, they will feed on fish, shellfish, small mammals, and the eggs and young of other birds. Gulls have been observed dropping clams and other shellfish on paved surfaces to break open the shells to feed on the soft interior.

4. The Middle Seashore

41. Where are mussels found? The edible mussel (*Mytilus edulis*) is widely distributed along the coasts of America and Europe and all the cool seas of the world. Forming closely crowded colonies on rocks, wharves, and almost any solid

Mussel

Byssus threads

object, these bivalves are considered a delicacy in Europe and are beginning to make inroads in this country. Mussels attach themselves by means of sticky byssus threads, which are produced by a gland.

42. What creature is described as "standing on its head and kicking food into its mouth with its feet"? The barnacle (*Balanus*) begins its life as a free-swimming larva that grows and molts several times until it reaches a stage at which it attaches to some solid object. It then builds the divided shell that

Barnacle

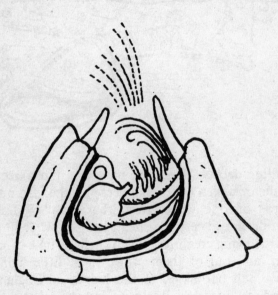

will be its home for the rest of its life. Because it uses its feathery feet to set up currents that drive plankton into its mouth and has attached itself by what could be considered the back of its neck, one zoologist has described the barnacle with the words quoted above.

43. How did gooseneck barnacles get that name? There are
two main kinds of barnacles: acorn and gooseneck. Acorn
barnacles are the familiar ones found on rocks and boat bottoms. Often washed up on the shore, gooseneck barnacles
(*Lepas*) usually are found attached to floating logs, buoys,

Gooseneck barnacle

and floating debris. They were named in the sixteenth cen-
tury in the mistaken belief that geese hatched from them. In
the words of one writer from that era, "Certaine trees
whereon do growe certaine shellfishes . . . wherein are con-
teined little living creatures; which shells in time of maturitie
do open, and out of them growe those little living foules
whom we call barnakles, in the North of England brant
geise, and in Lancashire tree geise, but the other that do fall
upon the land do perish and come to nothing."

**44. What are those yellowish strings of parchmentlike capsules
that we find on the beach?** These are the egg cases of the
knobbed or channeled whelks. They contain hundreds of
tiny, miniature whelks inside that would have hatched in
approximately two months if they had not been stranded on
the shore. They would have emerged through a round door in

Egg case of knobbed whelk

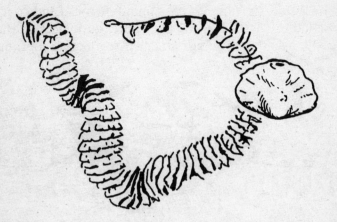

the wall of the capsule. The adult whelks are predators, preying on other mollusks, including clams and oysters, that it digs out of the bottom.

45. How does a jingle shell get that name? The jingle shell (*Anomia*) is a little yellowish-white, iridescent shell whose lower valve has a hole that allows the animal to anchor itself

Jingle shell

to rocks or any hard object. These shells make a jingling sound when the surf crashes upon the dead shells scattered on the beach; they make the same sound on a windy day if they are strung together.

46. What might you find at the tide line after high tide? The receding tide at the water's edge deposits many interesting

objects, some of which may have been carried great distances: driftwood, egg cases of the whelks, sand collars, mermaid's purses (the egg cases of the skate fish), and a variety of empty shells and crab parts.

47. Why is the periwinkle considered a transitional form? Some species of periwinkles (*Littorina*) seem to be evolving toward adaptation to a terrestrial life. Probably introduced from Europe, they have become quite common and can be

Periwinkle

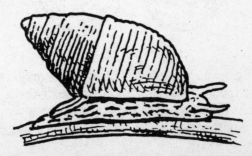

found on both coasts, on rocks and seaweed between tide limits. When the tide recedes, the periwinkles withdraw into their shells, closing their watertight operculums and retaining enough moisture to prevent them from drying out. In certain species they are able to live out of water for months, their gill chambers having been modified to act as lungs; in such cases they approach the status of true air breathers, such as land snails.

48. What are two of the most common kinds of seaweed? Sea lettuce (*Ulva lactuca*), which reaches a length of up to three feet, is common on both coasts. Anchored to rocks by its holdfast (see 49), this is the bright green algae that, when torn from its bond by storms, floats inshore and

Sea lettuce

Rockweed

washes up on the shore, causing an unpleasant odor. Rockweed (*Fucus vesiculosis*) is one of the world's most widely distributed seaweed. Olive green, with forked branches and swollen air bladders along a distinct midrib, rockweed masses serve as shelter for marine animals, preventing them from drying out.

49. What is a holdfast? Marine algae or seaweed are of two kinds, fixed and drifting. The latter are the diatoms, which number in the billions and play a critical role in the economy of the sea. Fixed seaweed, found along the shore and in

Holdfast

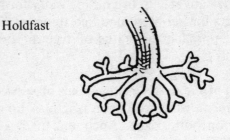

marine estuaries, are attached to rocks, the bottom, or some object by a holdfast. This resembles a root but differs from it in providing no nutrition; and though it branches like a root, its function is solely to anchor the seaweed.

50. What are the delicate, lacelike growths that are found on mussels and seaweed? Pick up a shell, a bit of seaweed, or a rock at low tide, and it is likely to have patches of various

Moss animals

colors on the surface. When examined with a hand lens or microscope, the patches transform into tiny colonies of spreading or treelike forms called moss animals (*Bryozoa*). The moss animals grow in two ways. One kind grows by encrusting, spreading in lacelike sheets on seaweed and other submerged objects. The other grows in a plantlike fashion and resembles seaweed. They feed on plankton, inhabit both shallow and deep water, and are easily confused with hydroids (see 74). Close examination discloses the difference: bryozoan tentacles are equipped with cilia for feeding, while hydroids possess nematocysts (see 73).

51. What are sand dollars? Sand dollars—also known as cake urchins or sea biscuits—are essentially flattened sea urchins. Their spines are much shorter, but they move about in a

Sand dollar

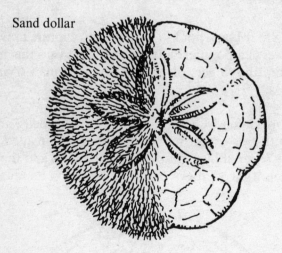

manner similar to that of sea urchins. They feed on diatoms and minute edible particles, passing them along with their tube feet to their mouth.

52. What is a beak thrower? The beak thrower (*Glycera*) is a bloodworm that is quite active. It burrows through the sand and is capable of shooting out its large proboscis, which is

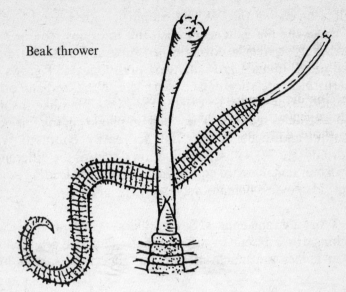

Beak thrower

armed with four black hooks that it uses to capture prey. The beak thrower is a popular saltwater bait and uses this mechanism to nip the fingers of an unwary fisherman trying to bait a hook.

53. What are chitons? Chitons, or coat-of-mail shells, are the most primitive of extant mollusks. With an elongated, flattened body covered by eight overlapping plates and the ability to attach itself to rocks or solid objects by its muscular

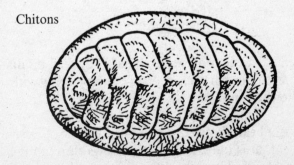

Chitons

foot, the chiton can withstand tides or enemies. Chitons thrive on a diet of diatoms, seaweed, and algae, which they rasp off with their radula. If detached, the chiton will curl up in the manner of a pill bug.

54. How did the ruddy turnstone acquire its name? This stocky, orange-legged shorebird (*Arenaria interpres*) waddles along the shore, using its bill to turn over shells, pebbles, seaweed,

Ruddy turnstone

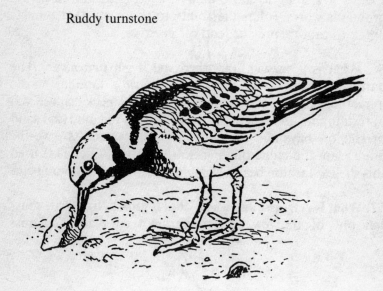

and driftwood in its search for its prey, the animals hiding underneath. Sometimes two birds will work together to move an object too heavy for one.

55. What is a ribbon worm? The ribbon worm is one of several kinds of marine worms. It can be found burrowing in muddy or sandy beaches. Highly elastic, it can extend to many times its normal length, with the largest American

Ribbon worm

species stretching to more than 75 feet. Also known as the proboscis worm, it has the ability to shoot out that tubular organ to capture prey or fend off enemies.

56. What is a quahog, and where did it get that name? The hard-shell clam (*Mercenaria mercenaria*), familiar in its immature stages as the cherrystone and littleneck clam served in restaurants on the half-shell, is found in the mud and sand of shallow bays and estuaries. The Narragansetts gave the adult clam the name quahog. Early Native Americans used this shell to fashion beads, stringing them into wampum belts.

57. What is eelgrass, and where does it grow? Not a true grass but one of the few totally marine seed plants, eelgrass

Eelgrass

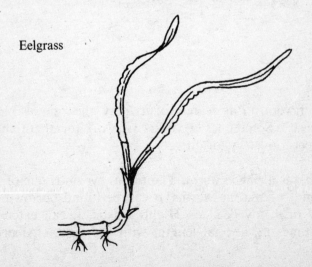

(*Zostera marina*) left the land millions of years ago to take up its watery abode. Growing in 2 to 6 feet of water on both coasts, it has a horizontally creeping stem rooted in sand or mud and long, thin, ribbonlike leaves. Its flowers, inconspicuous and short-lived, open underwater. It grows in sheltered, brackish estuaries, where it is an important provider of food and shelter for many small animals, fish, and invertebrates.

58. How does the moon snail move and feed? Equipped with a broad, muscular foot for mobility, the northern moon snail (*Lunatia heros*) digs into the sand, seeking its favorite food: a

Moon snail

clam, mussel, or other mollusk. When it finds it, the snail envelops its prey with its fleshy foot, drills a neat hole through the shell with its rasping tongue or radula, and consumes the innards.

59. What is a sand collar? The sand collar is the name of the circular egg mass that is pressed out from the side of the northern moon snail. It is composed of a sticky mucus material in which the eggs are imbedded. Since it is formed while the snail is buried in sand, it becomes covered with sand

Sand collar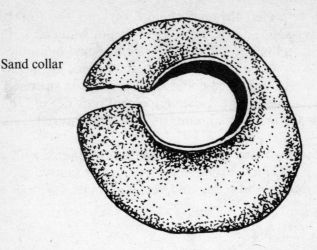

grains. When found on the shore, it has probably dried out and will crumble easily when handled.

60. Where is the common edible oyster found? Adult oysters require a specific range of salinity to thrive. These conditions are found in the shallow, brackish waters of bays and estuaries. There the female releases roughly 60 million eggs that develop into tiny, top-shaped larvae called trochophores. They swim about using their cilia or drift with the currents and soon settle to the bottom, where they will spend the rest of their lives.

61. What are the natural enemies of oysters? The oyster's major predators are the appropriately named oyster drill (*Urosalpinx cinerea*) and the common starfish (*Asterias*). Because of its high mortality in the early stages of life (it is often consumed by copepods and other crustaceans), the female oyster releases millions of eggs at a time.

62. How does the oystercatcher feed? The American oyster-catcher (*Haematopus palliatus*) may be found along the beach and mud flats at low tide, seeking to dine on clams, snails, and other crustaceans. Oysters are only a part of the

Oystercatcher

diet of this large bird, which uses its sturdy red bill to pry apart partially open bivalves and then neatly sever the adductor muscle (see 63) before the victim can close its shell.

63. What is an adductor? The adductor is a muscle in bivalves that draws the two shells together, protecting the soft inner parts. This adductor muscle is the edible part of the market scallop.

64. Are there starfish with more than five arms? Starfish (*Asteroidea*)—more properly called sea stars, since they are

Basket star

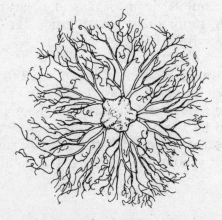

not fish—consist of a flat body and, typically, four or five radiating arms, although some have more (usually a multiple of five). For example, the sun star has ten arms, and the sunflower star has twenty. The basket star (*Gorgonocephalus arcticus*) has five arms with numerous branching tentacles, giving the appearance of a swarm of baby snakes.

65. Can starfish see? Not in the conventional way. They have eyespots at the tip of each ray. These light-sensitive organs allow them to react to various stimuli. Also on the tip are special tube feet that serve as organs of smell and touch.

66. How do starfish move? Starfish are equipped with an internal hydraulic system called the water vascular system. This circulatory system carries food and oxygen to the body and provides the means of movement. Branching from the water vascular system are hundreds of tiny tube feet in double rows, lining the grooves of the underside of the arms. By pumping water in and out of these tube feet, which are equipped with suction disks, the starfish is able to pull itself along or use these feet to open bivalves for food.

67. How do starfish feed? Most starfish are carnivorous and will feed on almost any kind of mollusk, worm, sea urchin, or even the young of their own species. Wrapping its arms around a clam, for example, it exerts an untiring pull on the two valves until its victim, exhausted, opens its shells. Then the starfish, turning its stomach inside out, surrounds the clam and slowly digests it.

68. What is a sea anemone? Often referred to as "flowers of the sea," sea anemones boast a stunning variety of colors, rivaling the finest floral display. They have a slitlike mouth surrounded by a fringe of stinging and/or sticky tentacles. When these tentacles contact their prey, the stinging cells discharge their venom into its body. The paralyzed victim is

Sea anemone

then moved to the mouth by contraction of the tentacles and swallowed. There are as many as one thousand species, ranging in width from a half-inch to 5 feet. They are found all over the world, from the intertidal zone to a depth of almost 6 miles.

69. What is an abalone? The abalone is a highly prized, edible snail occurring mainly on the Pacific coast. It hardly looks like a snail since its flattened spire is barely discernible. The abalone grazes on seaweed, clinging to rocks

Abalone

with a powerful foot. The inside of the shell of this animal is highly valued because of its beauty. Its iridescence has been likened to the effect of an oil film on water.

70. Can sea anemones move about? Related to corals, sea anemones belong to the class *Anthozoa* and are one of the few members of this class that are mobile. The anemone can extend portions of its base or pedal disk to pull itself along and move slowly from place to place.

71. What bird runs up and down following the breakers as they advance and recede on the shore? The sanderling (*Calidris alba*) occurs throughout the world, seemingly always in a

Sanderling

hurry, feeding at the very edge of the water, nimbly avoiding being overcome by incoming waves. Tiny crustaceans and mollusks, brought in by the waves and left stranded by the outgoing wave, are the sanderling's food source.

5. The Lower Seashore

72. What are cnidarians? The phylum *Cnidaria* (from the Greek *knide,* "nettle"), formerly known as *Coelenterata,* follows the phylum *Porifera* (sponges) in complexity of development. It includes hydroids, jellyfish, sea anemones, sea pens, and most corals. The *Cnidaria* usually take one of two shapes: one is the polyp, such as the sea anemone; the other is the medusa, as typified by the jellyfish. The polyp, radially symmetrical, is essentially a hollow tube or sac surrounded

Cnidarians

by a body wall with an opening, encircled with tentacles, that serves as a food inlet and an outlet for waste. The tentacles are armed with nematocysts (see 73) for feeding and protection. Some form colonies; others are free-swimming. The bell-shaped medusa displays a fringe of tentacles and swims by pulsating contractions. The mouth of the jellyfish is located on a stalk on the underside of the bell.

73. What is a nematocyst? All *Cnidaria* are equipped with batteries of stinging cells called nematocysts. These are microscopic capsules containing a coiled hollow thread, or tube, with a triggerlike projecting bristle. When some prey or predator comes in contact with the bristle, the tube is forcefully discharged. There are several kinds of nematocysts: those that penetrate and inject a poison, those that trap the intruder by entanglement, and those that are sticky like flypaper.

74. What are hydroids? On casual examination hydroids appear to be plants, but when viewed through a magnifying

Hydroids

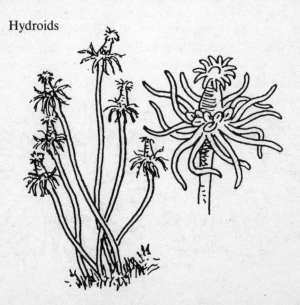

glass, they disclose their animal characteristics. The colony resembles a cluster of flowers, revealing many branches covered with tiny polyps, some of them displaying a tubelike mouth surrounded by waving tentacles that capture plankton to feed on. Other polyps are reproductive and release free-swimming medusae that will form new colonies. The dangerous Portuguese man-of-war is related to the lowly hydroid, since it is also a colony with trailing tentacles (40 or 50 feet long) that are equipped with stinging polyps attached to a balloonlike float that propels the colony by the wind.

75. Are horseshoe crabs true crabs? No. They are more closely related to land spiders and scorpions. One of the oldest living species on earth (there are fossil forms going back 200 million years), the horseshoe crab (*Limulus polyphemus*) is

Horseshoe crab

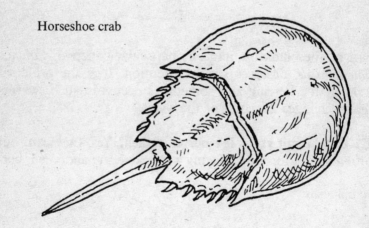

probably descended from the prehistoric trilobites that roamed those early seas. Also known as king crabs, they reach a length of 2 feet, including a tail spike. In late spring pairs may be seen onshore, where the females dig holes and then deposit their eggs. They are found on the East Coast, from Nova Scotia to Florida.

76. What is a copepod? Copepods, from the Greek words *kope* ("oar") and *pous* ("foot"), are tiny crustaceans, the most numerous multicellular animals in the world. They are

Copepod

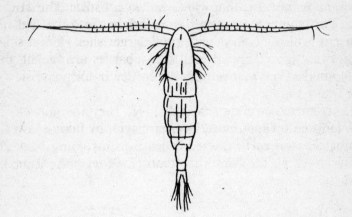

part of the plankton, and one particular copepod, *Calanus finmarchieus,* deserves special attention. It is the principal food of the herring and sometimes occurs in such swarms that it changes the color of the water.

77. Can hermit crabs live out of a shell? Yes, they can, but probably not for long. Having a soft, delicate abdomen, her-

Hermit crab

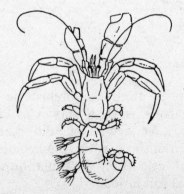

mit crabs (*Paguridae*) desperately need an empty shell to protect themselves from predators. Beautifully designed by nature, their bodies are curved in the proper counterclockwise direction that almost all snails are coiled, and are equipped with a hook on the end to grasp the inside of the shell. The front legs and claws are so constructed that they seal the opening when the crab withdraws into the shell. As the crab grows, it must seek out a larger shell (moon snails and whelks are favored), and many battles are fought with other hermit crabs, sometimes to the death, for possession of the desired home.

78. How do scallops move? Scallops swim by taking in water between their two shells and expelling it through one of the holes in the hinge, a dramatic example of jet propulsion. As they grow older, they are likely to attach themselves to a rock and remain there for the rest of their lives.

79. Can scallops see? Surprisingly, yes! Around the edge of the shell, projecting from the mantle, is a row of brilliant blue

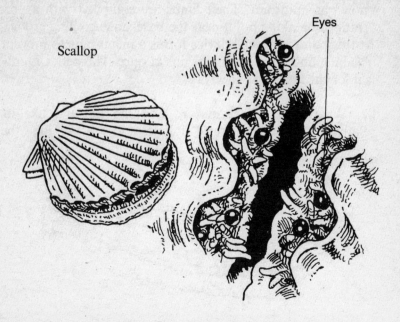

Scallop

Eyes

eyes, each consisting of a lens, a retina, and an optic nerve. If it espies an approaching predator, the scallop snaps shut its shell parts and shoots rapidly through the water.

80. Are sponges animals or plants? Sponges are simple, many-celled animals that come in many sizes, shapes, and colors. There are three main classes, which in some instances require microscopic examination for identification. Sponges are usually attached to the bottom of some solid object; some species encrust shells, plants, or rocks. Special cells draw water into the many openings of the sponge, bringing oxygen and minute food particles that sustain its growth. Other cells remove waste and carbon dioxide. Though sponges are sedentary, their eggs, after developing within the sponge, are ejected as tiny larvae covered with cilia that enable them to swim about for several hours or days and finally to settle on the bottom.

81. Can crabs regenerate legs and claws? Yes. Like many other sea creatures, crabs can grow a new leg or claw by a process known as autotomy. Their limbs are equipped with a pre-arranged breaking point near the base, an encircling groove. At this point, a special device forms a membrane to prevent bleeding, and regeneration begins at once. This reflex is critical for the creature's survival.

82. What is a sea cucumber? Looking nothing like its relatives—the starfish, sand dollars, and sea urchins—the

Sea cucumber

fat, leathery sea cucumber (*Cucumaria*) resembles that garden vegetable. It has the spiny skin and water vascular system typical of echinoderms. The mouth, at one end, is surrounded by ten branched, sticky tentacles that capture diatoms and are then brought into the mouth to be scraped clean. Asian species are considered a great delicacy.

83. Where are octopuses found? They are most often found inshore, though some species live in the deep sea. These agile and ferocious animals occur in nearly all seas of the world.

Octopus

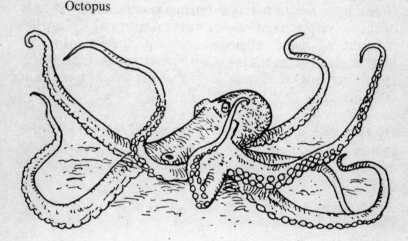

The common octopus (*Octopus vulgaris*) has an arm span of 4 feet and feeds by ambush and hunting. Although they differ in many respects from other members of the phylum *Mollusca,* octopuses do belong in this classification. Specifically, they belong to the class *Cephalopoda,* from the Greek *kephale* ("head") and *podus* ("foot").

84. How and why do the octopus and squid change color? Undisputed champions of color change in the animal world, octopuses and squid dazzle the eye with the speed and variety of colors they display to elude enemies or show emotions of anger or fear. Their skins have thousands of chromatophores containing the many colors, which are opened or closed by muscular action. The contraction and expansion of these cells varies the hues and their intensity, allowing these cephalopods to blend into the background and become nearly invisible at will.

85. What are the largest known kinds of squid and octopuses? The subject of myth and fiction, sea monsters have long fascinated followers of the sea. There are indeed huge squid; the largest confirmed specimen, *Architeuthis dux,* is almost 60 feet long, weighs nearly a ton, and is most likely the basis of the sea serpent tales. Sucker scars on sperm whales are the remnants of titanic undersea battles with these giant squid. Not nearly as large but impressive nevertheless is the North Pacific octopus, which reaches 15 feet and weighs more than 100 pounds.

86. How does the squid move in the water? With a torpedo-shaped body, the squid is well adapted to a free-swimming existence. Normally, the squid swims forward, using the

Squid

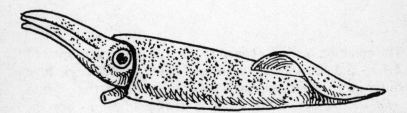

fins on the sides of its body in an undulating manner. However, water can be taken into the mantle cavity and powerfully ejected through its siphon, jet-propelling the squid at remarkably high speed. By directing the siphon, it can move backward or forward. If it is in danger, it can expel a cloud of dark ink to blind an attacker.

87. What is a shipworm? The shipworm (*Teredo*) is not really a worm, but a destructive bivalve mollusk. It starts life as a larva in the plankton and soon attaches itself to wharves, wooden pilings, and, in the days of wooden ships, to the bottoms of these vessels. In ancient times it destroyed Greek

Shipworm

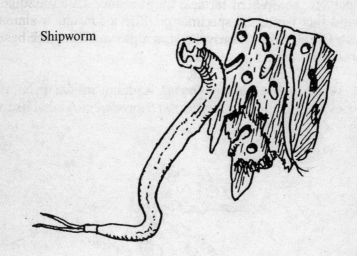

triremes and Roman galleys. Its body is long and wormlike, with the boring valves at one end and a pair of siphons on the other. The shipworm bores by twisting and turning, swallowing the shavings, extracting cellulose, and converting it into soluble sugar. Other food is obtained by the siphons: water is drawn in by one, filtered for small organisms, and then discharged through the other.

88. What is the lion's mane? The lion's mane (*Cyanea capillata*) is the largest jellyfish in the world, reaching a diameter of 8 feet, with tentacles 120 feet long. This, of course, is an extreme; the jellyfish does not ordinarily exceed 3 feet, and is usually about 1 foot in diameter. It reaches its maximum size in colder waters and has thousands of nematocysts, making it very dangerous to other sea creatures. Curiously, some young fish take advantage of the protection offered by this jellyfish and are permitted to seek haven beneath their host. The jellyfish makes no attempt to capture these fish.

89. Why is the pistol shrimp so named? There are several species of small lobsterlike shrimp, the largest of these being under 2 inches. When alarmed, they produce a loud snapping sound that is caused by a snapping device in one oversized claw. The effect is also used to stun a passing fish, making its capture easier.

90. Where are sea horses found? Ranging in size from 1.5 inches to 12 inches, the sea horse (*Hippocampus*) inhabits the

Sea horse

eelgrass and seaweed shallows. With an armored body and prehensile tail, this graceful fish is also unique in its reproductive behavior. The female deposits her eggs in a pouch on the male's belly, and he broods them. After about 6 weeks, a 100 or more perfect miniature sea horses are expelled and swim away to start their independent lives.

91. What are nudibranches? Also known as sea slugs, these beautifully decorated snails are shell-less mollusks. The decorations on their backs are in antlerlike, bushy projections that act as gills and serve for camouflage. The heads have tentacles and eyes. They feed on hydroids, barnacles, and

Nudibranch

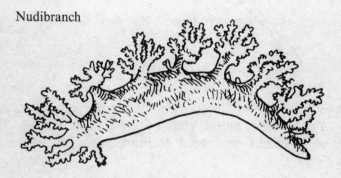

bryozoans. Interestingly, certain species of nudibranches can utilize the stinging cells of the hydroids they digest for their own use. Sea hares are members of this group. They have folds on their back, and the second pair of tentacles on their head resemble rabbit ears—hence their name. They graze on a variety of algae, absorbing the pigment from them and displaying those colors.

92. Can you write with a sea pen? Not really. Related to anemones but resembling a quill pen in form, sea pens range in size from about 4 inches to 6 feet. They are in the form of colonies with long stems and plumes made up of feeding

Sea pen

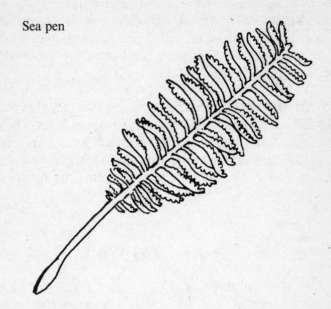

polyps. Sea pens are luminescent. When brought to the surface, a part of the plume, gently touched, will light up. The light will spread from branch to branch until the entire sea pen glows.

93. Why is the masking crab known by that name? One of the most unusual camouflage techniques is the one used by a West Coast spider crab sometimes known as a decorator crab (*Loxorhynchus crispatus*): it gathers bits of seaweed, sponges, and other matter and places them on barbed spines on its back. This collection of debris allows it to blend into its environment and elude its enemies. Other masking crabs have been seen swaying to simulate the movement of seaweed in a current of water.

94. What is a calico crab? The calico crab (*Ovalipes ocellatus*), also called lady crab, is a beautiful though pugnacious swim-

Calico crab

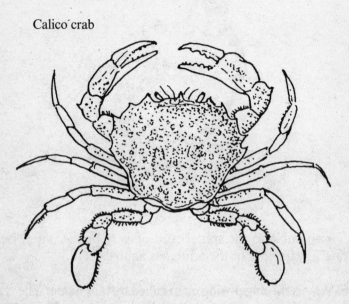

ming crab—delicate, pale, yellowish-gray, purple-spotted. Mobility comes from its hind legs, which propel it in modified paddlelike fashion. This crab burrows in sandy or mud bottoms at low tide, revealing only its eyes and antennae as it watches for prey or enemies. A good swimmer, it migrates to deeper water in winter.

95. What are sea squirts? Found attached to pilings, seaweed, or rocks, the sea squirt, a globular mass with a tough, flexible tunic, surprisingly belongs to the phylum *Chordata,* which includes the most highly developed animals on earth, including humans. This seemingly primitive-looking creature has two siphons close to the top of its domelike body. When touched, it contracts, forcefully expelling water from these openings—hence the name sea squirt. Scientists discovered the link to chordates by observing that the larvae, which are free-swimming, resemble tadpoles and have a notochord that

Sea squirt

is comparable to the spinal cord of vertebrates. This "back-bone" disappears in the adult sea squirt.

96. Why is the angel-wing clam called by that name? The angel wing (*Cyrtopleura costata*) is a burrowing clam inhabiting mud or clay. It is very rarely seen alive since it lives as much

Angel-wing clam

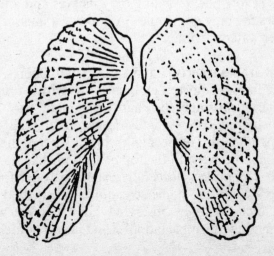

as 2 feet below the surface. The beautiful, fragile shells are found on the beach, and when cleaned and held together, the two shells resemble the white outspread wings of an angel. The presence of these clams is revealed by their extended large siphons, which retract in a flash when disturbed.

97. What is a mantis shrimp? Differing from most shrimps in having a flattened rather than a laterally compressed body, the mantis shrimp (*Squilla empusa*) has a pair of powerful

Mantis shrimp

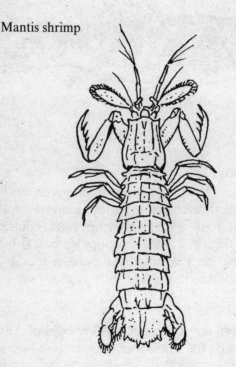

claws that resemble those of the praying mantis. This creature can extend those claws with remarkable speed to capture prey. Its pale yellowish-green body is 8 to 10 inches long. It lives in burrows in the mud, emerging at night to feed. Fishermen are extremely wary of the potential for injury when handling this creature.

98. What is a mummichog? The mummichog (*Fundulus heteroclitus*) is one of several species of killifish, commonly found in weedy areas close to shore. These creatures are useful to humans because they destroy the larvae of mosquitoes. They have been known to burrow to a depth of 6 inches into

Mummichog

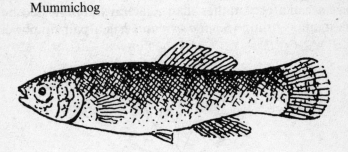

silt in the winter. The female is brownish-green, with darker bars on the side. Males are brighter: greenish-blue above, with the bars on the side displaying more contrast with the lighter body color. They are a popular bait fish because of their stamina on the hook. There is a closely related species, the California killifish (*Fundulus parvipinnis*), on the West Coast.

99. How large can the American lobster become? Allowed to grow undisturbed, the American lobster (*Homarus americanus*) can attain a weight of about 30 pounds. In fact, the heaviest lobster on record weighed 34 pounds, with a majority of the weight in its giant claws, the larger for crushing clams and mussels and the other, with its sharp teeth, for seizing and tearing up fish, carrion, or plant material. The lobster will feed on almost any dead animal it finds but will readily capture living fish or other marine animals. It can vary in color, but it is usually mottled, dark green above, yellow to orange beneath, and sometimes bright blue on the limbs. Lobsters live inshore during the summer and migrate to deeper water in winter.

100. Which crab is most sought commercially? Second only to the lobster, the blue crab (*Callinectes sapidus*) is a valuable crustacean that supports a large fishing industry. Like other crustaceans, this crab grows by molting, and before its new shell hardens, it is known as the familiar soft-shell crab. The

Blue crab

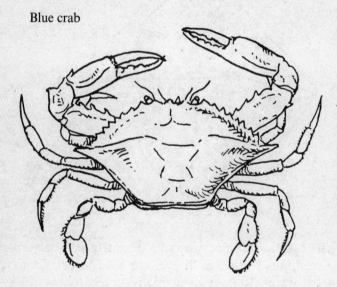

soft-shelled result is obtained by confining these crabs in boxes until they molt. This is another of the swimming crabs, with its hind legs flattened to form swimming organs. It is pugnacious and has great strength. The crab is fearless and will take on any opponent, including humans. Small barnacles are sometimes found attached to the shell. These crabs are found from Cape Cod to the Gulf Coast and are especially copious in Chesapeake Bay, which is famous for them.

101. What common inshore fish is related to sharks and the giant manta ray? It is the little skate (*Raja erinacea*), a brownish, flattened fish with large, winglike pectoral fins. The name comes from the Old Norse word *skata*. Skates lack true bones; instead, like sharks, their skeleton is made of cartilage. The little skate frequents shallow water on sandy bottoms. During daylight it lies partially buried in the sand, with

Little skate

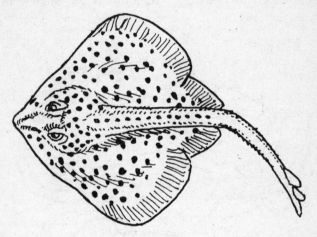

only its eyes and spiracles above the surface. The spiracles, large openings behind each eye, are used to extract oxygen from the water. It emerges at night to feed on a variety of crustaceans, including squid, clams, and worms, although it will take a fisherman's bait during the day. The skate produces the odd-shaped egg case that is sometimes found in the tide wrack and is known as mermaid's purse.